The
Seashore Life
of
Jersey

by

Marine Biology Section

of the

Société Jersiaise

SOCIÉTÉ JERSIAISE
2014

NEDBANK | PRIVATE WEALTH
SINCE 1834

The Seashore Life of Jersey

First published in Great Britain
in 2014 by Société Jersiaise
www.societe-jersiaise.org

ISBN 978 0 901897541

Contents

Introduction

This guide has been created to help those who find themselves on a Jersey beach looking at an animal or seaweed asking: *'I wonder what that is?'*

Every year the Société Jersiaise receives e-mails, letters and phone calls from people who have found interesting things on the beach which they cannot identify. Such enquires are always welcome (indeed we encourage them) but they highlight the need for a basic seashore guide which can introduce people to the wealth of marine life from around our coastline. We are therefore proud to present this illustrated guide to Jersey's seashore life.

Using the collective experience of the Marine Biology Section, we have photographed over 360 of the most commonly encountered seashore plants and animals on Jersey. This guide contains all these images together with some basic notes concerning names and habitats. We have also included a list of traditional Jersey-Norman French names which, although not much used these days, need to be recorded in print before they are lost entirely.

We hope that you will enjoy using this guide and that it will increase general awareness about the beauty and fragility of Jersey's marine environment. If you encounter any of the plants and animals featured in this guide then please let the Marine Biology Section know where and when you saw them. We need as much information as possible about our seashore life and are grateful to receive reports of all marine species, even the really common ones.

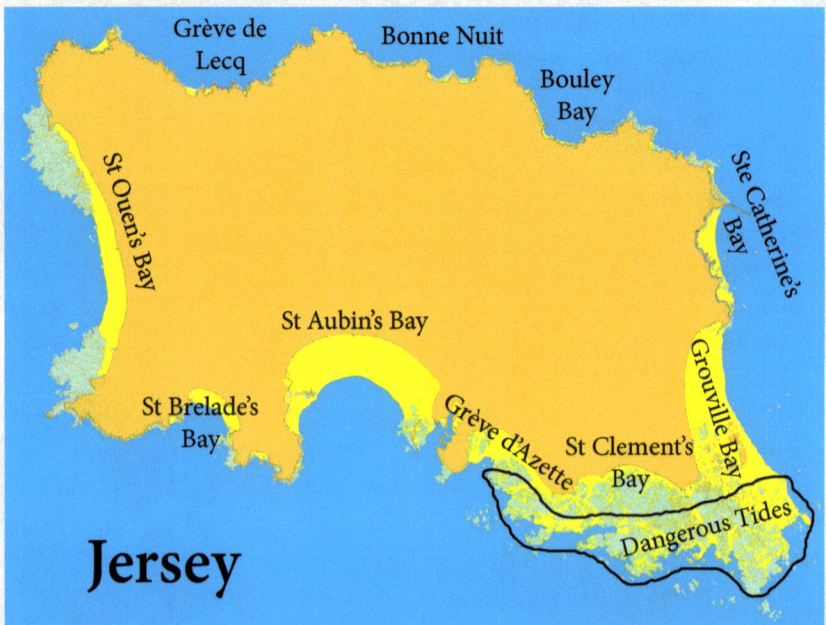

The island of Jersey at low water showing the main sandy and rocky seashore areas. Beware of the incoming tide, especially near headlands. It is recommended that people visit the area marked 'Dangerous Tides' only in the company of an experienced local guide.

About this Guide

Jersey is situated on the southern side of the English Channel where colder water species from the UK and Scandinavia overlap with warmer waters ones from southern Europe. As a consequence Jersey has a marine fauna and flora that is noticeably different to that of the United Kingdom.

Since 1949 the Marine Biology Section of the Société Jersiaise has been making practical studies of Jersey's seashore. This is no easy task as Jersey has a 12 metre tidal range which, at low water, can double the size of the island. Furthermore, Jersey's topography and geology creates a complex array of sandy and rocky habitats that support a diverse range of species. We have so far recorded over 3,200 different species from Jersey's marine environment but the true total for our shores is probably double this number.

This guide does not include detailed descriptions but the photographs provided should be clear enough to identify most species with reasonable confidence. If you need more detailed information then please consult one of the textbooks given in the Further Reading section (page 72); or you could get in touch with the Marine Biology Section and we will do our best to identify it for you.

If visiting Jersey's seashore then we request that you respect the island's fragile marine ecology and cause no disruption. In particular we ask that you:

- **Do not take or move animals or plants from their habitat.** If you cannot identify something then photograph it and contact us at the Société Jersiaise.

- **If you turn over a stone, always return it to its original position.** It takes over five years for a carelessly turned stone to regain all its marine life.

- **Dress appropriately.** Always wear warm, waterproof clothing and use boots or stout shoes. Carry a mobile phone and, if possible, a GPS or compass.

- **Consult a tide table.** Beware of Jersey's large tidal range, especially on the south and east coasts. NEVER linger on the lower shore after the tide has turned.

Take nothing but photographs. Leave nothing but your footprints.

The Marine Biology Section
Société Jersiaise, 7 Pier Road, St Helier, Jersey JE2 4XW
marinebiology@societe-jersiaise.org

Abbreviations Used in this Guide

✪ = *Occurs in the Channel Islands but is rare or absent from the United Kingdom.*

☻ = *Non-native and has been introduced into our region from foreign parts.*

☠ = *May sting, pinch or bite.* **DO NOT TOUCH!**

◼ = *Our records suggest that this species may be threatened and/or in decline locally.*

Upper/Mid/Lower Shore = *Denotes where on the shore the species can usually be found.*

Sand/Gravel/Rock/Seaweed/Pelagic = *The habitat where the species is most often found.*

Common/Frequent/Occasional/Rare = *How common the species is locally (estimated).*

(**+ XXX**) = *Describes the inset photograph, if it is a different species to the main picture.*

Hymeniacidon perlevis
Ragged Sponge - 30 cm
mid-lower shore; on rock; common

Halchondria panicea
Breadcrumb Sponge - 20 cm
mid-lower shore; on rock; common

Axinella dissimilis (+ out of water)
Yellow Staghorn Sponge - 20 cm
lower shore; on rock; occasional

Tethya citrina
Golf Ball Sponge - 5 cm
lower shore; on rock; frequent

Grantia compressa
Smooth Purse Sponge - 3 cm
lower shore; on seaweed; frequent

Sycon ciliatum
Hairy Purse Sponge - 2 cm
lower shore; on seaweed; frequent

Sponges

Polymastia penicillus (+ *P. boletiformis*)
Chimney Sponge - 15 cm
lower shore; on rock; frequent

Dysidea fragilis
Goosebump Sponge - 20 cm
lower shore; on rock; frequent

Ophlitaspongia papilla
Red Crater Sponge - 10 cm
lower shore; on rock; occasional

Ciocalypta penicillus
Spire Sponge - 10 cm
lower shore; on sand/rock; occasional

Terpios gelatinosa
Purple Sponge - 5 cm
mid-lower shore; on rock; common

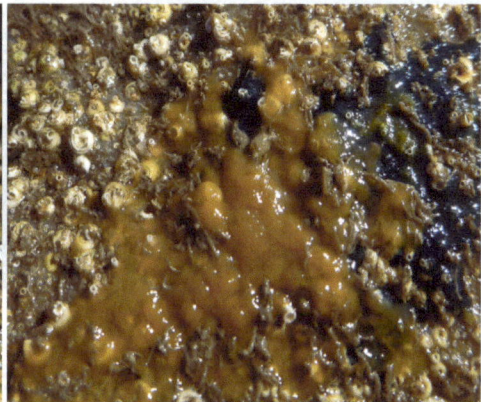

Halisarca dujardinii
Blobby Sponge - 20 cm
mid-lower shore; on rock; common

Anemonia viridis 🦑
Snakelocks Anemone - 15 cm
mid-lower shore; on rock; common

Cereus pedunculatus
Daisy Anemone - 15 cm
mid-lower shore; on sand; common

Actinia fragacea
Strawberry Anemone - 10 cm
lower shore; on rock; occasional

Actinia equina
Beadlet Anemone - 5 cm
upper-lower shore; on rock; common

Urticina felina
Dahlia Anemone - 20 cm
lower shore; on rock; frequent

Aulactinia verrucosa
Gem Anemone - 5 cm
upper-lower shore; on rock; frequent

Calliactis parasitica
Parasitic Anemone - 10 cm
lower shore; on rock/shells; occasional

Adamsia carciniopados
Cloak Anemone - 5 cm
lower shore; with Hermit Crabs; rare

Aiptasia mutabilis
Trumpet Anemone - 5 cm
lower shore; on rock; occasional

Actinothoe sphyrodeta
Fried Egg Anemone - 5 cm
lower shore; shaded rocks; occasional

Caryophyllia smithii
Devonshire Cup Coral - 2 cm
lower shore; on rock; occasional

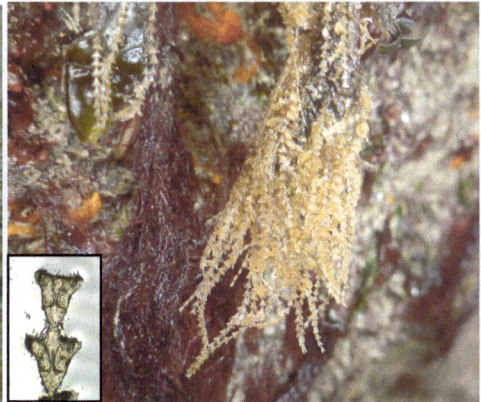

Dynamena pumila
Hydroid - 10 cm
lower shore; on rock/seaweed; frequent

Aurelia aurita 🦂 (+ *Cyanea lamarckii* 🦂)
Moon Jellyfish - 35 cm
pelagic; free-swimming; frequent

Chrysaora hysoscella 🦂
Compass Jellyfish - 30 cm
pelagic; free-swimming; occasional

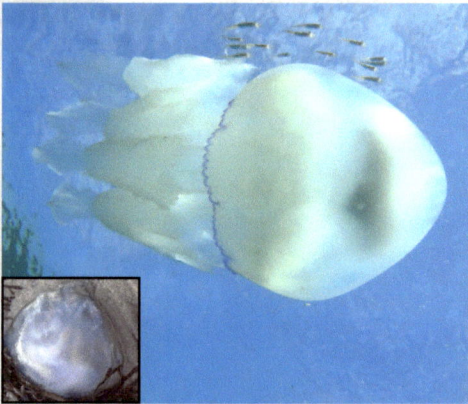

Rhizostoma pulmo 🦂
Barrel Jellyfish - 1.2 metres
pelagic; free-swimming; frequent

Pleurobranchia pileus (+ *Beroe cucumis*)
Sea Gooseberry - 15 cm
pelagic; free-swimming; frequent

Symsagittifera roscoffensis ✪
Mint-sauce Worm - 0.5 cm
mid shore; on gravel; common

Prostheceraeus vittatus
Candy-striped Flatworm - 5 cm
lower shore; on rock; occasional

Harmothoe extenuata
Scaleworm - 4 cm
lower shore; on rock; occasional

Pelogenia arenosa ✪ ⬛
Giant Scaleworm - 25 cm
lower shore; in gravel; rare

Sabella pavonina
Peacock Worm - 15 cm
lower shore; in sand; common

Sabella spallanzanii ✪
Feather-duster Worm - 35 cm
lower shore; marinas/rockpools; rare

Megalomma vesiculosum
Gravel Worm - 5 cm
lower shore; in gravel; frequent

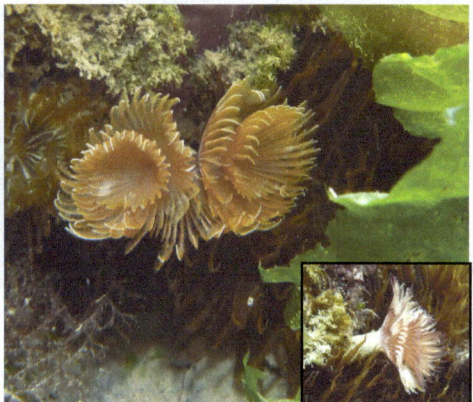

Bispira volutacornis
Double Spiral Worm - 10 cm
lower shore; on rock; occasional

Lanice conchilega
Sandmason Worm - 5 cm
mid-lower shore; in sand; common

Owenia fusiformis
Crusty Worm - 6 cm
lower shore; in sand; frequent

Chaetopterus variopedatus (+ whole tube)
Parchment Worm - 25 cm
lower shore; in sand; frequent

Spirobranchus lamarcki
Keelworm - 3 cm
mid-lower shore; on rock; common

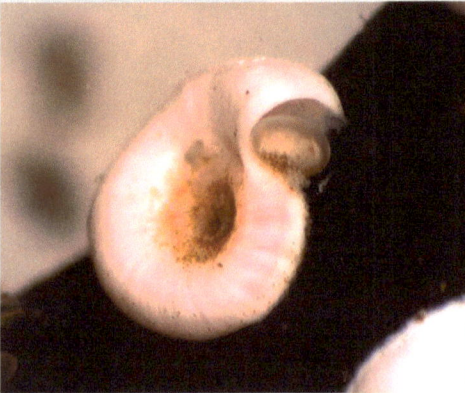

Spirorbis spirorbis
Spiral Worm - 0.5 cm
mid-lower shore; on rock; common

Filograna implexa
Spaghetti Worm - 3 cm
lower shore; on rock; frequent

Bristle Worms

Arenicola defodiens
Black Lugworm - 20 cm
lower shore; in sand; occasional

Arenicola marina (+ sand cast)
Blow Lugworm - 12 cm
mid-lower shore; in sand; common

Nereis sp. 🐾
Ragworm - 25 cm
lower shore; in sand; common

Nephtys sp.
Catworm - 25 cm
mid-lower shore; in sand; common

Phyllodoce lamelligera
Large Paddleworm - 60 cm
lower shore; in sand; frequent

Eteone picta
Painted Paddleworm - 6 cm
lower shore; in sand; occasional

Sabellaria spinulosa
Ross Worm - 10 cm
lower shore; on rock; frequent

Egg of an annelid worm - 3 cm
mid-lower shore; in sand; common

Marphysa sanguinea 🐛
Verm - 60 cm
lower shore; in gravel; occasional

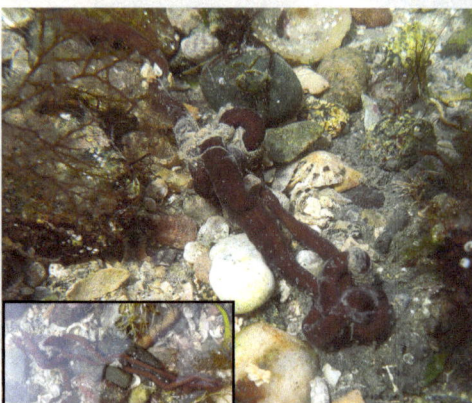

Lineus longissimus
Bootlace Worm - 10 metres+
lower shore; under rocks; occasional

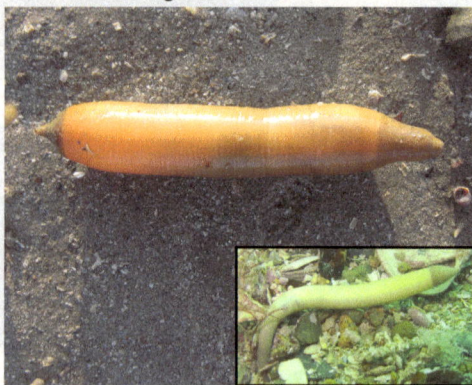

Golfingia vulgaris
Fat Spoon Worm - 15 cm
lower shore; in gravel; common

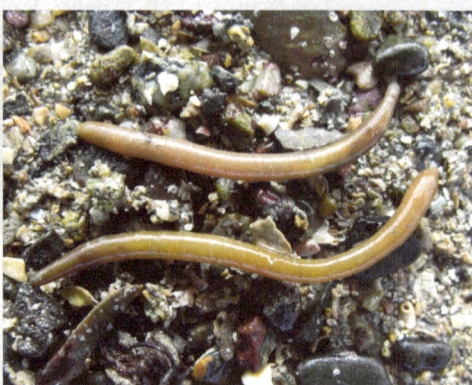

Golfingia elongata
Thin Spoon Worm - 10 cm
lower shore; in gravel; common

Chthamalus montagui
Diamond Barnacle - 0.5 cm
upper shore; on rock; common

Austrominius modestus 👽
Australian Barnacle - 0.5 cm
upper-lower shore; on rock; frequent

Semibalanus balanoides
Acorn Barnacle - 0.5 cm
mid-lower shore; on rock; common

Verruca stroemia
Twisted Barnacle - 1 cm
lower shore; on rock; frequent

Perforatus perforatus
Volcano Barnacle - 1 cm
mid-lower shore; on rock; common

Lepas anatifera
Goose Barnacle - 10 cm
offshore; on driftwood; occasional

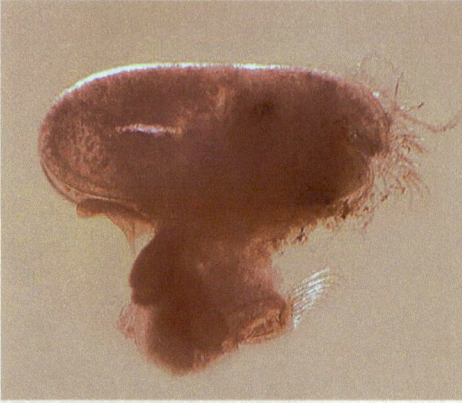

Prionotoleberis norvegica ✪
Norwegian Ostracod - 0.2 cm
lower shore; in sand; occasional

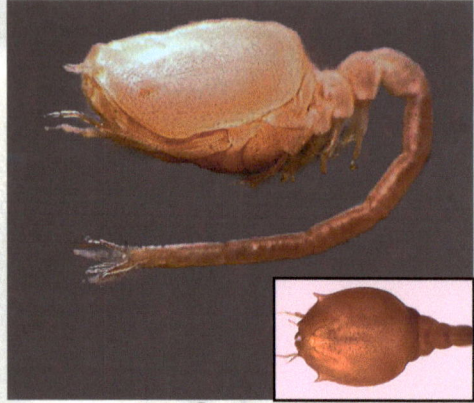

Eocuma dollfusi ✪ 🖼
Devil Shrimp - 0.3 cm
mid-lower shore; on seagrass; rare

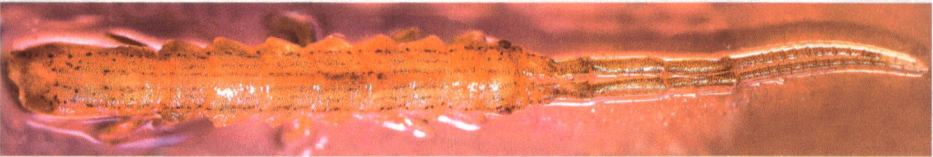

Idotea linearis
Stick Isopod - 4 cm
lower shore; on seaweed; occasional

Idotea emarginata
Emarginated Isopod - 3 cm
lower shore; on seaweed; rare

Idotea balthica
Baltic Isopod - 3 cm
lower shore; on seaweed; frequent

Sphaeroma serratum
Beach Woodlouse - 1 cm
mid-lower shore; under rocks; frequent

Anilocra frontalis 🦟
Fish Louse - 4 cm
pelagic; parasitic on fish; common

Ligia oceanica
Sea Slater - 5 cm
upper shore; on rock; common

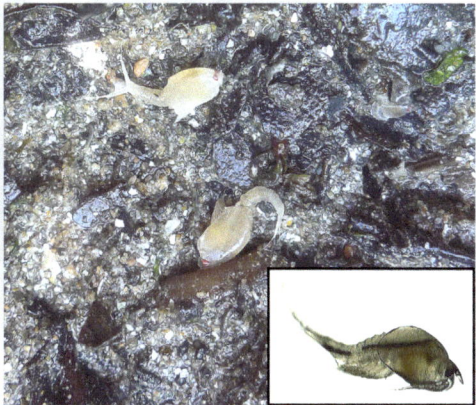

Nebalia bipes
Leptostracan Shrimp - 0.5 cm
lower shore; under rocks; occasional

Talitrus saltator
Sand-hopper - 2 cm
upper shore; under rocks; common

***Gammarus* sp.**
Shore Hopper - 1.5 cm
mid-lower shore; under rocks; common

Crangon crangon
Common Shrimp - 9 cm
lower shore; on sand; common

Palaemon serratus
Common Prawn - 10 cm
mid-lower shore; rockpools; common

Athanas nitescens
Silver-backed Prawn - 2 cm
lower shore; under rocks; frequent

Periclimenes sagittifer ✪
Anemone Prawn - 3 cm
offshore; on anemones; occasional

Eualus occultus (+ Hippolyte varians)
Chamaeleon Prawn - 2 cm
lower shore; under rocks; frequent

Upogebia deltaura
Burrowing Prawn - 15 cm
lower shore; in gravel; occasional

Pestarella tyrrhena
White Burrowing Prawn - 7 cm
lower shore; in sand; occasional

Axius stirynchus
Lipotte - 15 cm
lower shore; under rocks; occasional

Galathea squamifera
Brown Squat Lobster - 4 cm
mid-lower shore; under rocks; common

Galathea strigosa 🦐
Painted Squat Lobster - 10 cm
lower shore; under rocks; occasional

Homarus gammarus 🦐
Lobster - 60 cm
lower shore; under rocks; occasional

Eupagurus bernhardus 🦐
Soldier Hermit Crab - 10 cm
mid-lower shore; on sand; frequent

Pagurus prideaux
Cloaked Hermit Crab - 10 cm
lower shore; on sand; frequent

Diogenes pugilator
Sinistral Crab - 2 cm
mid-lower shore; on sand; common

Porcellana platycheles
Broad-clawed Porcelain Crab - 2 cm
mid-lower shore; under rocks; common

Pisidia longicornis
Long-clawed Porcelain Crab - 1 cm
lower shore; under rocks; common

Inachus dorsettensis
Spider Crab - 8 cm
lower shore; on rock; occasional

Macropodia deflexa (+ M. tenuirostris)
Spider Crab - 12 cm
lower shore; on rock; occasional

Pisa tetraodon (+ Pisa armata)
Four-horned Spider Crab - 15 cm
lower shore; on seaweed; frequent

Ebalia tumefacta
Nut Crab - 3 cm
lower shore; on sand; rare

Maja squinado 🦀
Edible Spider Crab - 40 cm
lower shore; on sand; frequent

Carapace Only

Cancer pagurus 🦀
Chancre - 30 cm
mid-lower shore; under rocks; common

Atelecyclus rotundatus
Circular Crab - 10 cm
lower shore; in gravel; rare

Thia scutellata 🗻
Thumbnail Crab - 5 cm
lower shore; in gravel; rare

Corystes cassivelaunus
Masked Crab - 15 cm
lower shore; in sand; occasional

Pirimela denticulata
Toothed Primela - 5 cm
lower shore; in sand; occasional

Necora puber 🦀
Lady Crab - 20 cm
lower shore; under rocks; frequent

Liocarcinus navigator
Navigator Crab - 15 cm
lower shore; in sand; occasional

Liocarcinus depurator
Harbour Crab - 10 cm
lower shore; in sand; frequent

Carcinus maenas 🦀
Green Crab; Shore Crab - 12 cm
lower shore; on rock/sand; common

Pilumnus hirtellus
Hairy Crab - 6 cm
lower shore; under rocks; common

Xantho pilipes ◣
Furrowed Crab - 15 cm
lower shore; under rocks; occasional

Hemigrapsus sanguineus 👽
Asian Crab - 10 cm
mid-lower shore; under rocks; occasional

Nymphon gracile
Sea Spider - 8 cm
mid-lower shore; on seaweed; occasional

Pyrrhocoris apterus
Clown-faced bug - 2 cm
extreme upper shore; on soil; occasional

Anurida maritima
Rockpool Springtail - 0.5 cm
upper shore; rockpools; occasional

Hydroschendyla submarina
Beach Centipede - 5 cm
upper shore; under rocks; frequent

Antalis vulgaris
Tusk Shell - 5 cm
offshore; in sand; occasionally washed up

Lepidochitona cinerea
Grey Chiton - 3 cm
lower shore; on rock; common

Leptochiton scabridus ✪
Pink-footed Chiton - 0.8 cm
lower shore; under rocks; rare

Acanthochitona fascicularis (+ plate detail)
Large Bristly Chiton - 6 cm
lower shore; under rocks; common

Acanthochitona crinitus (+ plate detail)
Small Bristly Chiton - 4 cm
mid-lower shore; under rocks; common

Haliotis tuberculata ✪ ▨
Ormer - 10 cm
lower shore; under rocks; occasional

Diodora graeca
Keyhole Limpet - 3 cm
lower shore; under rocks; frequent

Patella vulgata
Common Limpet - 5 cm
upper-lower shore; on rock; common

Patella ulyssiponensis
China Limpet - 3 cm
lower shore; on rock; frequent

Patella depressa
Black-footed Limpet - 3 cm
mid-lower shore; rockpools; common

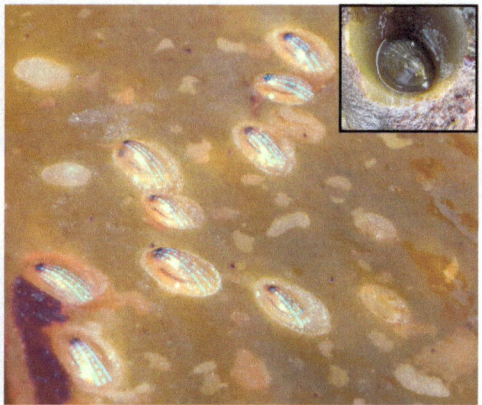

Helcion pellucidum
Blue-rayed Limpet - 2 cm
lower shore; on *Laminaria*; common

Osilinus lineatus
Toothed Top Shell - 2.5 cm
upper-mid shore; on rock; common

Gibbula magus
Turban Top Shell - 2.5 cm
lower shore; on sand; common

Calliostoma zizyphinum
Painted Top Shell - 3 cm
lower shore; on rock; common

Gibbula cineraria
Grey Top Shell - 1.5 cm
lower shore; under rocks; common

Gibbula umbilicalis
Purple Top Shell - 2 cm
mid-lower shore; on rock; common

Gibbula pennanti ✪
Jersey Top Shell - 1.5 cm
lower shore; on rock; common

Littorina obtusata
Flat Winkle - 1 cm
mid-lower shore; on seaweed; common

Littorina mariae
Spire Winkle - 1 cm
mid-lower shore; on seaweed; common

Molluscs (Gastropods)

Littorina littorea
Black Winkle - 3 cm
mid-lower shore; on rock; frequent

Littorina saxatilis
Rough Winkle - 1 cm
upper-mid shore; on rock; common

Melarhaphe neritoides
Small Periwinkle - 0.5 cm
upper shore; on rock; occasional

Cingula trifasciata
Banded Snail - 0.5 cm
upper-mid shore; under rocks; frequent

Rissoa parva (+ Pusillina sarsi)
Comma Shell - 0.5 cm
mid-lower shore; on seaweed; common

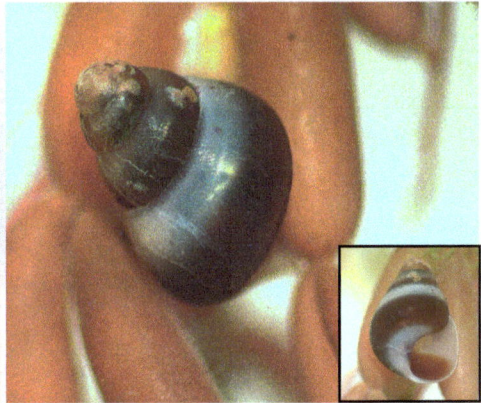

Barleeia unifasciata
Barlee's Sea Snail - 0.5 cm
lower shore; on seaweed; frequent

Trivia arctica
European Cowrie - 1 cm
lower shore; under rocks; common

Trivia monacha
Spotted Cowrie - 1 cm
lower shore; under rocks; frequent

Polinices catenus (+ egg ribbon)
Necklace Shell - 3.5 cm
lower shore; on sand; occasional

Polinices polianus (+ egg ribbon)
Alder's Necklace Shell - 2 cm
lower shore; on sand; rare

Lamellaria latens (+ *Lamellaria perspicua*)
Transparent Lamellaria - 3 cm
lower shore; under rocks; occasional

Crepidula fornicata ☻
American Slipper Limpet - 5 cm
lower shore; on rock/sand; common

Calyptraea chinensis
Chinaman's Hat - 1.5 cm
lower shore; on rock/shell; frequent

Epitonium clathrus
Common Wentletrap - 4 cm
lower shore; on sand; occasional

Nucella lapillus (+ *N. lapillus* f. *imbricata*)
Common Dog Whelk - 4 cm
mid-lower shore; on rock; common

Eggs of the Common Dog Whelk - 0.5 cm

Ocinebrina aciculata ✪
Small Sting Winkle - 1.5 cm
lower shore; under rocks; common

Ocenebra erinaceus
European Sting Winkle - 6 cm
lower shore; on rock; common

Hinia reticulata
Netted Dog Whelk - 3.5 cm
mid-lower shore; on sand; common

Eggs of Netted Dog Whelk - 0.5 cm

Buccinum undatum ◩ (+ egg mass)
Common Whelk - 10 cm
lower shore; on rock/gravel; occasional

Berthella plumula
Smooth Sea Lemon - 6 cm
lower shore; under rocks; frequent

Aplysia punctata (+ egg ribbon)
Sea Hare - 20 cm
mid-lower shore; on sand; frequent

Elysia viridis (+ egg ribbon)
Green Elysia - 1.5 cm
mid-lower shore; on *Codium*; frequent

Molluscs (Gastropods, Sea Hares, Sea Slugs)

Doris pseudoargus (+ egg ribbon)
Warty Sea Lemon - 12 cm
lower shore; on rock; frequent

Aeolidia papillosa (+ *Polycera quadrilineata*)
Grey Sea Slug - 12 cm
lower shore; on rock; occasional

Aeolidiella alderi
Collared Sea Slug - 2 cm
mid-lower shore; under rocks; occasional

Onchidella celtica
Irish Sea Slug - 2 cm
mid-lower shore; on rock; rare

Myosotella myosotis
Crevice Snail - 1.2 cm
upper shore; under rocks; occasional

Truncatella subcylindrica ◼
Looping Snail - 0.4 cm
upper shore; under rocks; rare

31

Glycymeris glycymeris
Dog Cockle - 7 cm
lower shore; in gravel; common

Mytilus edulis
Edible Mussel - 5 cm
mid-lower shore; on rock; occasional

Modiolus barbatus
Bearded Mussel - 10 cm
lower shore; on rock; rare

Modiolus phaseolina
Bean Mussel - 2 cm
lower shore; on rock; occasional

Pecten maximus
Great Scallop - 15 cm
lower shore; on sand; occasional

Mimachlamys varia (+ sponge *Suberites ficus*)
Variegated Scallop - 6 cm
mid-lower shore; under rocks; common

Molluscs (Bivalves)

Ostrea edulis
Native Oyster - 15 cm
lower shore; on rock; occasional

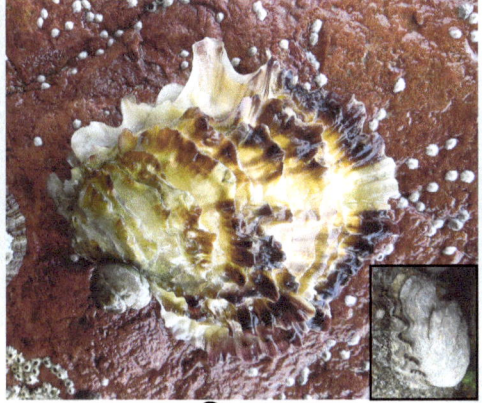

Crassostrea gigas
Pacific Oyster - 20 cm
mid-lower shore; on rock; common

Cerastoderma edule
Common Cockle - 5 cm
mid-lower shore; in sand; common

Acanthocardia tuberculata
Rough Cockle - 10 cm
lower shore; in sand; rare

Laevicardium crassum
Norwegian Cockle - 8 cm
lower shore; in gravel; common

Loripes lucinalis
Milky Shell - 3 cm
mid-lower shore; in sand; frequent

Lasaea rubra
Red Crevice Shell - 0.5 cm
upper-mid shore; under rocks; frequent

Spisula solida
Surf Clam - 5 cm
lower shore; in sand; common

Venus verrucosa
Praire - 8 cm
lower shore; in gravel; frequent

Clausinella fasciata
Banded Venus - 4 cm
lower shore; in gravel; frequent

Macoma balthica
Baltic Tellin - 3 cm
lower shore; in sand; frequent

Arcopagia crassa
Blunt Tellin - 6 cm
lower shore; in gravel; occasional

Molluscs (Bivalves)

Polititapes rhomboides
Banded Paloudre - 6 cm
lower shore; in gravel; common

Venerupis corrugata
Pullet Paloudre - 5 cm
lower shore; in gravel; occasional

Ruditapes decussatus
Chequered Paloudre - 8 cm
mid-lower shore; in gravel; frequent

Ruditapes philippinarum 👽
Manila Clam - 8 cm
mid-lower shore; in gravel; frequent

Tapes aurea
Golden Paloudre - 5 cm
lower shore; in gravel; occasional

Dosinia exoleta
Rayed Artemis - 6 cm
lower shore; in gravel; occasional

Abra alba
Mud Clam - 2 cm
lower shore; In Mud/Sand; frequent

Pandora inaequivalvis
Pandora Shell - 3 cm
lower shore; in sand; occasional

Lutraria angustior
Otter Shell - 15 cm
lower shore; in gravel; frequent

Mactra glauca ✪ ◪
Five-shilling Shell - 12 cm
lower shore; in sand; occasional

Gari fervensis
Faroe Sunset Shell - 5 cm
lower shore; in sand; rare

Gari depressa
Large Sunset Shell - 7 cm
lower shore; in gravel; occasional

Donax variegatus ✪
Flattened Donax - 4 cm
lower shore; in sand/gravel; frequent

Scrobicularia plana
Peppery Furrow Shell - 6 cm
mid shore; in mud; occasional

Thracia villosiuscula
Smooth Thracia - 4 cm
lower shore; in sand; occasional

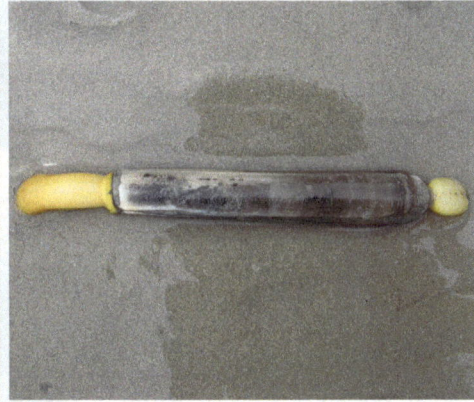

Solen Marginatus
Straight Razor - 12 cm
lower shore; in sand; common

Ensis ensis
Curved Razor - 12 cm
lower shore; in sand/gravel; frequent

Ensis arcuatus
Bandy Razor - 15 cm
lower shore; in gravel; common

Sepiola atlantica
Little Cuttle - 2 cm
lower shore; in sand; occasional

Sepia officinalis
Common Cuttlefish - 40 cm
offshore; on sand; common

Schizoporella unicornis
Unicorn Sea Mat - 8 cm
lower shore; on seaweed; common

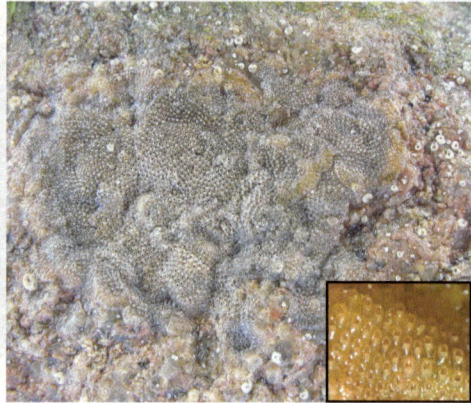

Cryptosula pallasiana
Black-eyed Sea Mat - 10 cm
mid-lower shore; under rocks; common

Escharoides coccinea
Domed Orange Sea Mat - 3 cm
mid-lower shore; under rocks; common

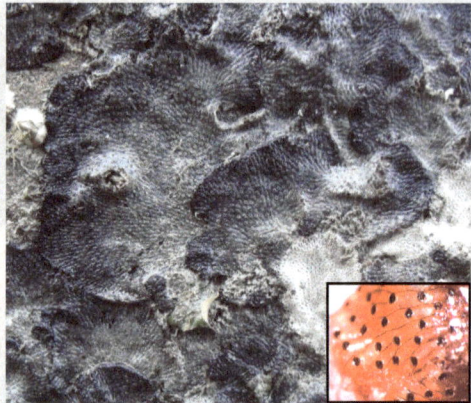

Watersipora subtorquata 👽
Asian Sea Mat - 50 cm
mid-lower shore; on rock; common

Asterina gibbosa (+ *A. phylactica*)
Cushion Star - 5 cm
mid-lower shore; under rocks; common

Marthasterias glacialis
Spiny Starfish - 70 cm
lower shore; under rocks; rare

Ophiura albida
White Brittlestar - 5 cm
lower shore; in sand; occasional

Ophiothrix fragilis
Common Brittlestar - 10 cm
lower shore; under rocks; occasional

Amphipholis squamata
Small Brittlestar - 2 cm
mid-lower shore; under rocks; common

Amphiura brachiata
Long-armed Brittlestar - 15 cm
lower shore; in sand; rare

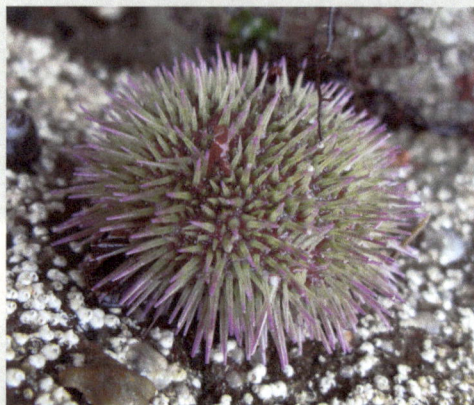

Psammechinus miliaris 🦟 📉
Purple-spined Urchin - 5 cm
lower shore; under rocks; occasional

Spatangus purpureus
Purple Heart Urchin - 12 cm
lower shore; in gravel; rare

Pawsonia saxicola
White Sea Cucumber - 5 cm
lower shore; under rocks; rare

Leptosynapta cruenta ✪/*L. galliennii* ✪
Burrowing Sea Cucumber - 20 cm
lower shore; in sand; occasional

Ciona intestinalis
Common Sea Squirt - 15 cm
lower shore; on rock; frequent

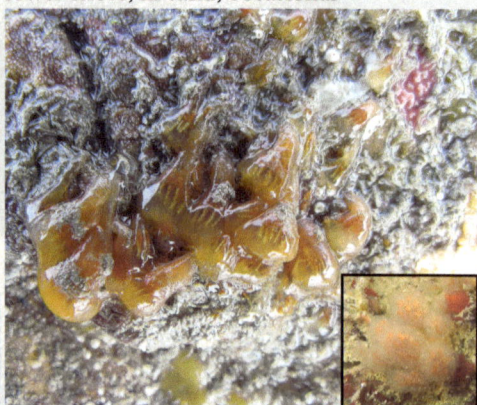

Morchellium argus
Flat-topped Sea Squirt - 4 cm
lower shore; under rocks; occasional

Pycnoclaella aurilucens (+ *Clavelina lepadiformis*)
Light Bulb Sea Squirt - 2 cm
lower shore; under rocks; occasional

Ascidia mentula
Red Sea Squirt - 16 cm
lower shore; on rock; frequent

Mogula tubifera
Horned Sea Squirt - 3 cm
lower shore; under rocks; occasional

Styela clava 👽
Leathery Sea Squirt - 12 cm
lower shore; on rock; frequent

Botryllus schlosseri
Star Ascidian - 20 cm
mid-lower shore; on rock; common

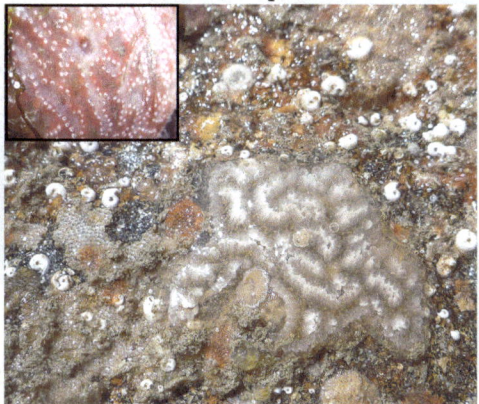

Botrylloides leachii
Chain Ascidian - 12 cm
lower shore; on rock; occasional

Lissoclinum perforatum
Perforated Sea Squirt - 8 cm
lower shore; under rocks; frequent

Phallusia mammillata
Lumpy Sea Squirt - 15 cm
lower shore; on rock; frequent

Dendrodoa grossularia
Gooseberry Sea Squirt - 2 cm
lower shore; on rock; occasional

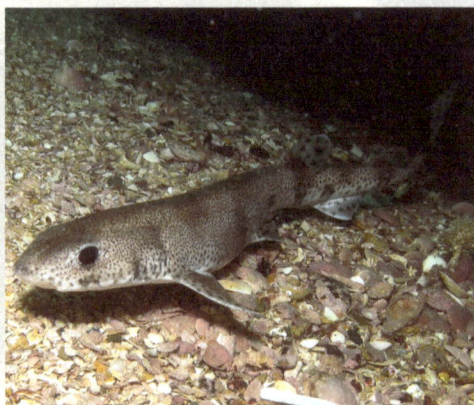

Scyliorhinus canicula
Lesser Spotted Catshark (Dogfish) - 1 metre
lower shore-offshore; on sand; frequent

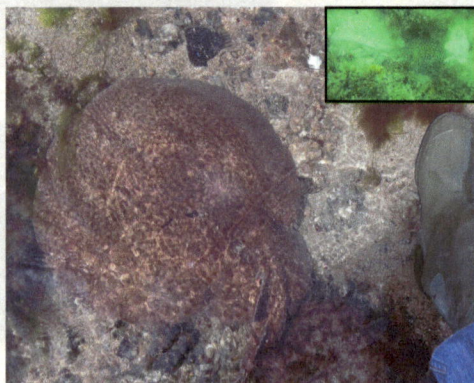

Torpedo marmorata ✪ 🐛
Marbled Electric Ray - 60 cm
offshore; on sand/rock; occasional

Raja undulata ✪
Undulate Ray - 1 metre
offshore; on sand/rock; occasional

42

Sea Squirts and Fish

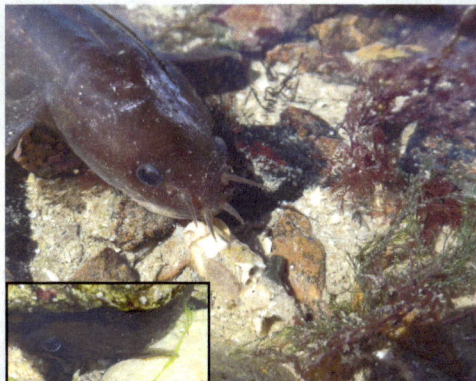

Ciliata mustela
Five-bearded Rockling - 25 cm
lower shore; under rocks; occasional

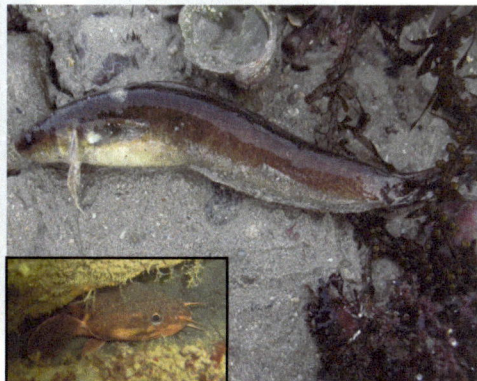

Gaidropsarus mediterraneus
Shore Rockling - 40 cm
lower shore; under rocks; frequent

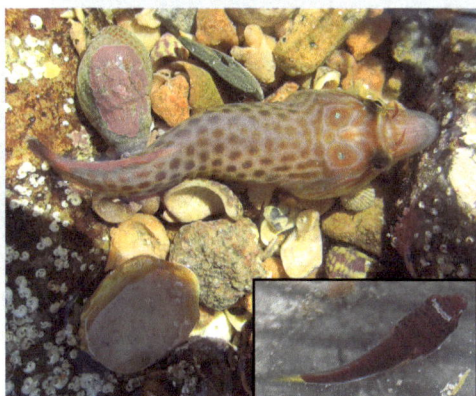

Lepadogaster lepadogaster (+ *Apletodon* sp.)
Shore Clingfish - 5 cm
lower shore; under rocks; common

Conger conger 🦴
Conger Eel - 2 metres
lower shore; among rocks; occasional

Zeus faber
John Dory - 90 cm
offshore; pelagic; occasional

Mola mola
Ocean Sunfish - 2.5 metres
offshore; pelagic; occasional

43

Syngnathus rostellatus
Nilsson's Pipefish - 20 cm
offshore; on seaweed; occasional

Nerophis lumbriciformis
Worm Pipefish - 10 cm
lower shore; under rocks; frequent

Hippocampus hippocampus ◩
Short-snouted Seahorse - 15 cm
offshore; on seaweed; rare

Spinachia spinachia
Sea Stickleback - 20 cm
lower shore; rockpools; rare

Taurulus bubalis
Long-spined Sea Scorpion - 20 cm
lower shore; rockpools; occasional

Cyclopterus lumpus
Lumpsucker - 60 cm
lower shore; under rocks; rare

Fish

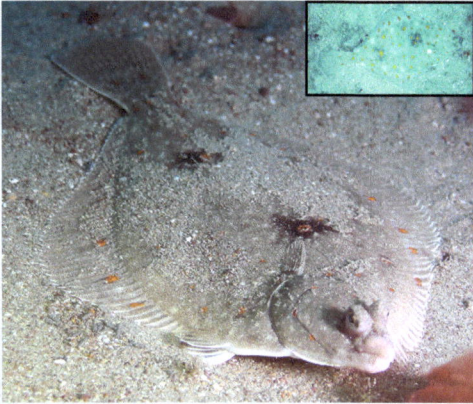

Pleuronectes platessa
Plaice - 90 cm
lower shore; on sand; occasional

Zeugopterus punctatus
Topknot - 25 cm
lower shore; under rocks; occasional

Solea solea
Common Sole - 70 cm
lower shore; on sand; occasional

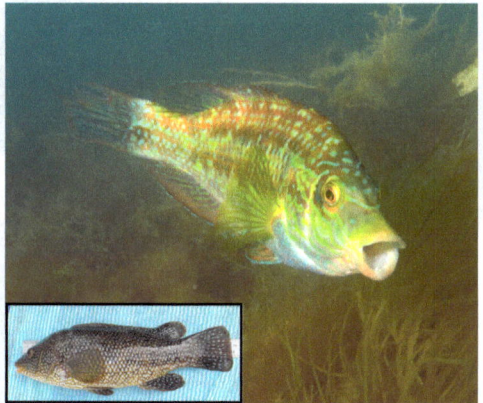

Symphodus melops (+ *Labrus bergylta*)
Corkwing Wrasse - 15 cm (+ Ballan Wrasse)
lower shore; rockpools; occasional

Echiichthys vipera
Lesser Weever - 15 cm
lower shore; in sand; occasional

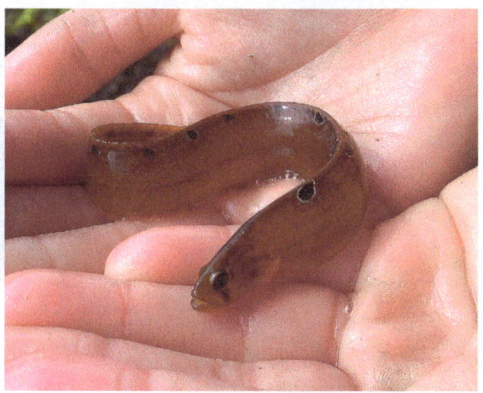

Pholis gunnellus
Butterfish / Rock Gunnel - 20 cm
lower shore; rockpools; occasional

Gobius paganellus
Rock Goby - 12 cm
lower shore; rockpools; frequent

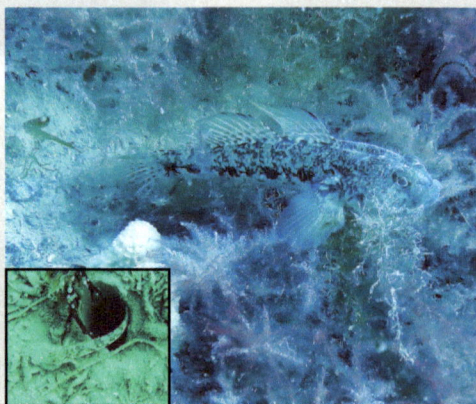

Pomatoschistus pictus
Painted Goby - 3 cm
lower shore; on sand; common

Thorogobius ephippiatus
Leopard-spotted Goby - 13 cm
lower shore; rockpools; rare

Gobius cobitis ✪ ▨
Giant Goby - 28 cm
lower shore; rockpools; rare

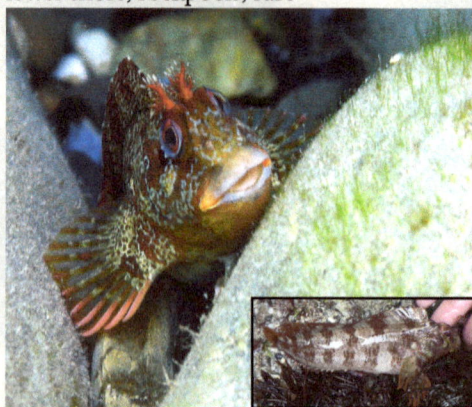

Parablennius gattorugine
Tompot Blenny - 30 cm
lower shore; under rocks; occasional

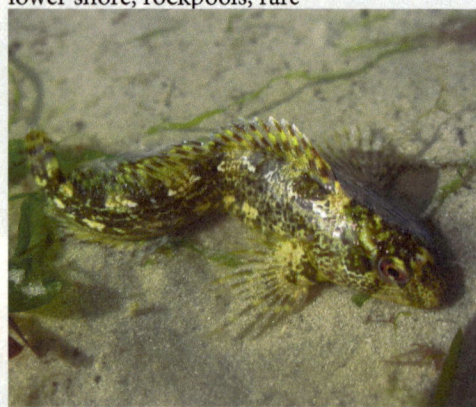

Lipophrys pholis 🐾
Shanny - 25 cm
lower shore; under rocks; common

Callionymus reticulatus 🦀
Reticulated Dragonet - 10 cm
lower shore; on sand; occasional

Dicentrarchus labrax 🐟 (+ *Atherina presbyter*)
Bass - 25 cm
lower shore; under rocks; occasional

Balistes capriscus
Triggerfish - 60 cm
offshore; pelagic; occasionally washed up

Ammodytes sp.
Lesser Sandeel - 20 cm
lower shore; in sand; common

Tursiops truncatus
Bottlenose Dolphin - 4 metres
offshore; pelagic; frequent

Halichoerus grypus
Grey Seal - 3 metres
offshore; pelagic; frequent

Rivularia bullata
A cyanobacterium - 3 cm
mid-lower shore; on rock; frequent

Verrucaria maura
Black Tar Lichen - 15 cm
upper shore; on rock; common

Verrucaria mucosa
Smooth Tar Lichen - 20 cm
mid-lower shore; on rock; common

Lichina pygmaea
Tufted Shore Lichen - 25 cm
upper shore; on rock; common

Ramalina subfariacea
Sea Ivory - 5 cm
extreme upper shore; on rock; common

Xanthoria parietina
Sunburst Lichen - 10 cm
extreme upper shore; on rock; common

Lichens and Red Seaweeds

Lithophyllum sp.
Encrusting Red Seaweed - 20 cm
mid-lower shore; on rock; common

Corallina officinalis
Sea Moss - 8 cm
mid-lower shore; rockpools; common

Jania rubens
Slender Coral Weed - 5 cm
lower shore; on rock; frequent

Mesophyllum lichenoides
Pink Plates - 2 cm
lower shore; on *Corallina*; frequent

Calliblepharis jubata
False Eyelash Weed - 30 cm
lower shore; on rock; common

Calliblepharis ciliata
Beautiful Eyelash Weed - 30 cm
lower shore; on rock; occasional

Catenella caespitosa
Creeping Chain Weed - 2 cm
upper shore; on shaded rock; frequent

Chondracanthus acicularis
Creephorn - 10 cm
mid-lower shore; on rock; common

Dumontia contorta
Dumont's Tubular Weed - 25 cm
mid-lower shore; rockpools; common

Furcellaria lumbricalis
Clawed Fork Weed - 30 cm
lower shore; on rock; common

Gracilaria bursa-pastoris ✪
Shepherd's Purse - 30 cm
lower shore; on rock; occasional

Rhodothamniella floridula
Sandbinder - 3 cm
lower shore; on rock; common

Halopithys incurva
Red Sea Pine - 15 cm
lower shore; on rock; frequent

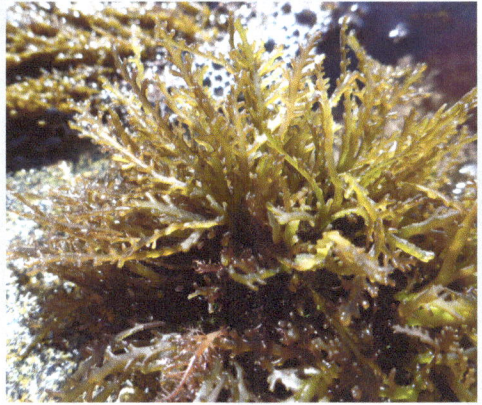

Osmundea osmunda
Royal Fern Weed - 20 cm
lower shore; rockpools; frequent

Osmundea pinnatifida
Pepper Dulse - 8 cm
mid-lower shore; on rock; common

Osmundea hybrida
False Pepper Dulse - 5 cm
mid-lower shore; on rock; frequent

Laurencia obtusa
Brittle Fern Weed - 15 cm
lower shore; rockpools; occasional

Asparagopsis armata 👽
Harpoon Weed - 30 cm
lower shore; rockpools; frequent

Ahnfeltiopsis devoniensis
Devonshire Fan Weed - 10 cm
lower shore; rockpools; occasional

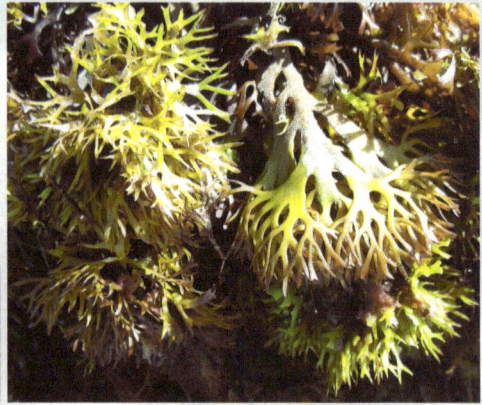

Chondrus crispus
Irish Moss - 20 cm
mid-lower shore; on rock; common

Mastocarpus stellatus
Grape-pip Weed - 15 cm
lower shore; on rock; common

Cryptopleura ramosa
Crinkle Weed - 25 cm
lower shore; on rock; occasional

Delesseria sanguinea
Sea Beech - 25 cm
lower shore; rockpools; rare

Gastroclonium ovatum
Red Grape Weed - 15 cm
lower shore; rockpools; frequent

Red Seaweeds

Gastroclonium reflexum
Reflexed Grape Weed - 5 cm
mid-lower shore; on rock; frequent

Lomentaria articulata
Bunny Ears - 10 cm
lower shore; on rock; frequent

Ceramium sp.
Banded Claw Weed - 25 cm
mid-lower shore; rockpools; common

Polysiphonia sp.
Siphon Weed - 30 cm
mid-lower shore; rockpools; common

Polyopes lancifolius 🛸 ✪
Furry Oyster Weed - 80 cm
mid-lower shore; rockpools; rare

Grateloupia subpectinata 🛸
Pacific Fringe Weed - 90 cm
mid-lower shore; rockpools; frequent

Palmaria palmata
Dulse - 50 cm
lower shore; On Rock/Seaweeds; common

Nemalion helminthoides
Sea Noodle - 40 cm
lower shore; on rock; occasional

Porphyra purpurea
Purple Laver - 70 cm
mid-lower shore; on rock; common

Porphyra dioica
Black Laver - 50 cm
upper shore; on rock; frequent

Meredithia microphylla
Mermaid's Ear - 3 cm
lower shore; on rock; occasional

Boergeseniella fruticulosa
Tufted Shrub Weed - 20 cm
mid-lower shore; rockpools; common

Vertebrata lanosa (on *Ascophyllum*)
Parasitic Weed - 5 cm
mid-lower shore; on seaweed; common

Desmarestia aculeata
Landlady's Wig - 8 metres+
lower shore; rockpools; occasional

Laminaria digitata
Oarweed - 1.5 metres
lower shore; on rock; common

Saccorhiza polyschides
Furbellows - 2 metres
lower shore; on rock; frequent

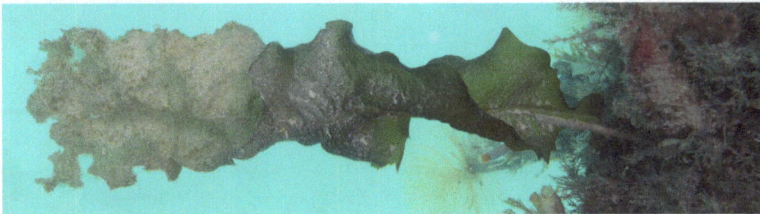

Undaria pinnatifida (above) 👽
Wakame - 2 metres
lower shore; rockpools/marinas; occasional

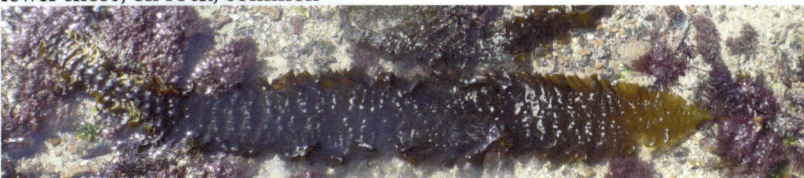

Saccharina latissima (below)
Sugar Kelp - 1.5 metres
lower shore; on rock; common

Halidrys siliquosa
Sea Oak - 1.5 metres
lower shore; rockpools; frequent

Sargassum muticum 👽
Wireweed - 8 metres
mid-lower shore; rockpools; common

Chorda filum
Bootlace Weed - 7 metres
lower shore; in sand; common

Himanthalia elongata
Thong Weed - 1.3 metres
lower shore; on rock; common

Ascophyllum nodosum
Egg Wrack - 1.5 metres
mid shore; on rock; common

Pelvetia canaliculata
Channel Wrack - 15 cm
upper shore; on rock; common

Brown Seaweeds

Fucus spiralis
Spiral Wrack - 25 cm
upper-mid shore; on rock; common

Fucus vesiculosus
Bladder Wrack - 70 cm
mid shore; on rock; common

Fucus serratus
Serrated Wrack - 60 cm
mid-lower shore; on rock; common

Bifurcaria bifurcata
Tuning-fork Weed - 50 cm
mid-lower shore; rockpools; common

Asperococcus bullosus
Fat Sausage Weed - 20 cm
mid-lower shore; rockpools; frequent

Pylaiella littoralis
Brown Filament Weed - 30 cm
mid-lower shore; on seaweed; common

Cystoseira nodicaulis
Bushy Noduled Weed - 70 cm
lower shore; rockpools; frequent

Cystoseira foeniculacea
Bushy Feather Weed - 90 cm
lower shore; rockpools; occasional

Cystoseira tamariscifolia
Bushy Rainbow Weed - 1 metre
lower shore; rockpools; frequent

Cystoseira baccata
Bushy Berry Weed - 1.5 metres
lower shore; rockpools; frequent

Dictyopteris polypodioides
Netted Wing Weed - 30 cm
lower shore; on rock; frequent

Padina pavonica
Peacock's Tail - 10 cm
mid-lower shore; on rock; rare

Petalonia fascia
Rockpool Weed - 35 cm
lower shore; rockpools; common

Cladostephus spongiosus
Hairy Sand Weed - 30 cm
lower shore; on rock; common

Stypocaulon scoparium
Sea Flax - 20 cm
lower shore; rockpools; frequent

Colpomenia peregrina
Oyster Thief - 10 cm
mid-lower shore; rockpools; common

Chordaria flagelliformis
Slimy Whip Weed - 50 cm
lower shore; rockpools; frequent

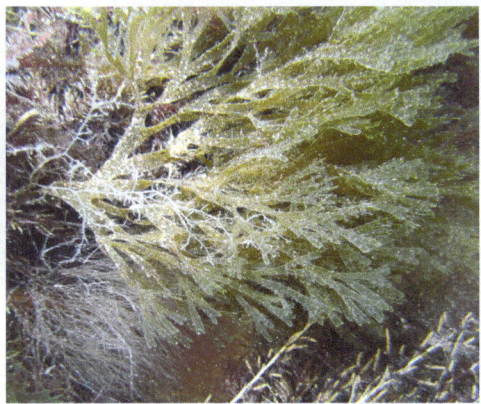

Dictyota dichotoma
Brown Fan Weed - 30 cm
lower shore; rockpools; common

Cladophora rupestris
Common Branched Weed - 20 cm
mid-lower shore; on rock; common

Acrosiphonia arcta
Tarantula Weed - 25 cm
mid-lower shore; rockpools; frequent

Ulva lactuca
Sea Lettuce - 30 cm
mid-lower shore; on rock; common

Ulva intestinalis
Gut Weed - 40 cm
upper-lower shore; on rock; common

Ulva compressa
Gutter Weed - 25 cm
mid-lower shore; on gravel; common

Blidingia sp.
Outfall Weed - 10 cm
upper-mid shore; on rock; common

Codium tomentosum (+ microscope view)
Native Velvet Horn - 30 cm
mid-lower shore; rockpools; frequent

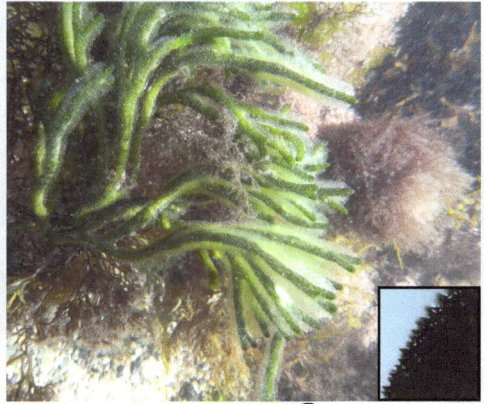

Codium fragile var. *fragile* 🛸
Non-native Velvet Horn - 30 cm
mid-lower shore; rockpools; occasional

Chaetomorpha linum
Flax Brick Weed - 15 cm
mid-lower shore; on gravel; frequent

Prasiola stipitata
Guano Weed - 2 cm
extreme upper shore; on guano/rock; rare

Zostera marina 🏴
Seagrass - 2 metres
lower shore; on sand; common

Zostera noltii 🏴
Dwarf Seagrass - 25 cm
mid-lower shore; on sand; common

Strand Line Finds

Scyliorhinus canicula
Mermaid's Purse - 6 cm
eggcase: Lesser Dogfish
common

Scyliorhinus stellaris
Mermaid's Purse - 12 cm
eggcase: Nursehound
rare

Raja brachyura
Mermaid's Purse - 12 cm
eggcase: Blonde Ray
common

Raja undulata
Mermaid's Purse - 8 cm
eggcase: Undulate Ray
common

Raja microocellata
Mermaid's Purse - 8 cm
eggcase: Small-eye Ray
occasional

Raja clavata
Mermaid's Purse - 7cm
eggcase: Thornback Ray
occasional

Strand Line Finds

Sepia officinalis
cuttlebone: Cuttlefish - 25 cm
common

Sepia orbignyana
cuttlebone: Rose Cuttlefish - 11 cm
common

Flustra foliacea
Hornwrack - 15 cm
common

Physalia physalis 😨
Portuegese-Man-o-War - 3 metres+
rarely stranded (very dangerous)

Buccinum undatum
Eggball of Common Whelk - 15 cm
common

Janthina janthina
Violet Sea Snail - 4 cm
rare

Salsola kali
Prickly Seawort - 50 cm
sandy shores; uncommon

Beta vulgaris subsp. ***maritima***
Sea Beet - 40 cm
rocky shores and shingle; common

Atriplex portulacoides
Sea Purslane - 90 cm
sand and shingle; occasional

Armeria maritima
Sea Thrift - 20 cm
rocky shores; occasional

Spergularia rupicola
Rock Sea-spurrey - 25 cm
rocky shores; frequent

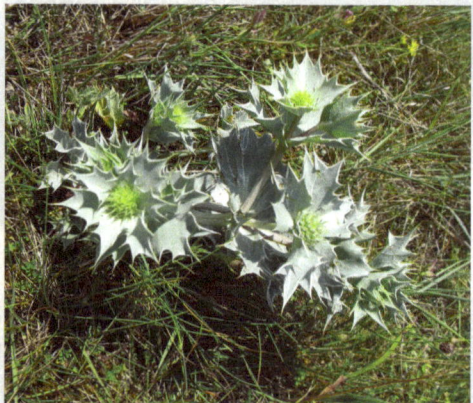

Eryngium maritimum
Sea-holly - 50 cm
sandy shores; occasional

Jèrriais Species Names

Below is a list of species names in Jèrriais, Jersey's native language. Notes concerning the context of the words are given in square brackets. This list was initially compiled by L'Office du Jèrriais as part of their *Dictionnaithe Jèrriais-Angliais* (Le Don Balleine, 2008). It was subsequently checked and expanded by Paul Chambers using a variety of sources including works by Frank Le Maistre, Ronald Le Sueur and Joseph Sinel as well as several kind people who were prepared to share their knowledge of Jèrriais words and phrases.

Sponges

êponge (f) en boursicot
Sycon ciliatum; purse sponge

êponge (f) en jaune dé
Axinella dissimilis; yellow staghorn sponge

Jellyfish & Anemones

leune (f)
Rhizostoma pulmo; barrel jellyfish

dgèrryi (m) Portûndgais
Physalia physalis; Portuguese man o' war

sangsue (f); con (m) d'Vênus; con (m) d'rotchi
sea anemones (general term)

oeu (m) d'putain
Anemonia viridis; snakelocks anemone

pîssenliet (m) d'mé
Metridium senile; plumose anemone

Worms

souothis (f) d'mé
Aphrodita aculeata; sea-mouse

**brun sandron (m);
sandon [St Ouen, La Rocque, Gorey];
sangdon[St Clement, Grouville, St Martin]**
Arenicola marina; common lugworm

nier sandron (m)
Arenicola defodiens; black lug

vèr (m) dé rocque; vèr (m) dé rotchi
Marphysa sanguinea; rock-worm

Arthropods

pêtre (m) à pipet
Nymphon gracile; sea spider

bênaque (f); bênèque (f); bânnêque (f)
barnacles (general)

p'tit gris baîni (m)
Semibalanus balanoides; acorn barnacle

bênacl'ye (f)
Lepas anatifera; goose barnacle

chèrvîn (m)
Schistomysis spiritus; fairy shrimp

poux d' mé (m)
Ligia oceanica; sea slater

puche dé grève (f)
Orchestia gammarellus; shore-hopper

saûticot (m)
Talitrus saltator; sand hopper

grôsse chèrvette (f)
Palaemon serratus; common prawn

blianche chèrvette (f)
Pestarella tyrrhena; burrowing prawn

grise chèrvette (f)
Crangon crangon; shrimp

êcrévette (f); langoustinne (f)
Nephrops norvegicus; langoustine

lipotte (f)
Axius stirhynchus; burrowing prawn

honmard (m); crotchet (m) [under-size];
quart (m) [small]; d'mi (m) [medium];
dé priy (m) [large]; chouque (f) [v. large];
couotheux (m) [v. large];
mouaithe (f) [female with eggs];
tréhar (m) [male]
Homarus gammarus; lobster

crabe (f) à co
Palinurus elephas; crawfish

êcrelle (f); êcrévisse (f)
Scyllarus arctus; slipper lobster

hèrmites (mpl)
Hermit crabs (general term)

lorme (f);
soudard (m) [Grouville]
Pagurus bernhardus; soldier crab

êcrévisse (f)
Galathea spp.; squat lobster

démouaîselle (f);
d'mouaîselle (f)
Porcellana platycheles; porcelain crab
Pisidia longicornis; porcelain crab
Pilumnus hirtellus; hairy crab
Thia scutellata; thumbnail crab
Pirimela denticulata; toothed pirimela

crabes (fpl)
Crabs (general term)

dormeuse (f)
Dromia personata; sleeping crab

pihangne (f);
hueûlîn (m) [St Ouen];
moussu (m) [juvenile];
mousseuse (f) [juvenile];
Maja squinado; spider crab

tréhar (m)
Macropodia spp.; spider crab
Inachus spp.; spider crab

petite pihangne (f);
trapenard (m) [old fashioned term]
Pisa armata; Gibb's spider crab
Pisa tetradon; four-horned spider crab

crabe (f); chancre (m); poingclios (m);
houais (m); ouais (m)
Cancer pagurus; chancre crab

rouoge crabe (f); vielle (f); vielle crabe (f)
Xantho pilipes; furrowed crab

crabe (f) grégeaise; démouaîselle (f);
d'mouaîselle (f); grégeaise (f); grégeon (m)
Necora puber; lady crab

couôrrêsse (f); fouaitheuse (f);
vèrte crabe (f); d'mouaîselles (fpl) [small]
Carcinus maenas; shore crab

Molluscs

ormèr (m)
Haliotis tuberculata; ormer

d'mouaîselle (f)
Littorina obtusata; flat periwinkle
Littorina mariae; flat periwinkle

vlicot sauvage (m)
Littorina saxatilis; rough periwinkle

bliu vlicot (m)
Gibbula umbilicalis; flat top shell
Gibbula pennanti; Jersey top shell

châté (m); topie (f) [St Ouen];
vlicot d'Espangne (m)
Calliostoma zizyphinium; painted top shell

colîn (m); chorchi (m); chorchiéthe (f);
vricq (m); gris vlicot (m)
Osilinus lineatus; toothed top shell

baîni (m);
êcaillard (m) [large]
Patella vulgata; limpet

néthe coque (f); nièr vlicot (m); coque (f);
fliée (f); vlicot (m) sauvage
Littorina littorea; black winkle

vlique (f) dé chance
Trivia arctica; cowrie

cône (f)
Charonia lampas; triton

colînmachon d'me (m);
colînmachon d'bâsse-ieau (m)
Euspira catena; necklace shell

églyise (f); vlique (f) d'églyise
Epitonium clathrus; wentletrap

bênarde (f); chorchiéthe (f);
coque (f) à j'vaux; vlique (f) chorchiéthe;
blianche coque (f)
Nucella lapillus; dog whelk

coque (f); coqueluche (f); vlique (f)
vlique (f) suaïse [small]
Buccinum undatum; whelk

linmache dé mé (f) [St Ouen];
sordonne (f) [Grouville]
Aplysia punctata; sea hare

lînmache (f) dé mé à rouoges pitchets;
sordonne (f) à rouoges pitchets;
rouoge-brînge (f); rouoge-pitchet
Coryphella browni; sea slug

suchette (f) [Grouville];
bobbe (f) [Grouville];
chuchette (f) [St Ouen]
Glycymeris glycymeris; dog cockle

becque-dé-corbîn (f); bé (m) d'corbîn;
orté (m) dé geniche; chuchette (f)
moûle (f)
Mytilus edulis; edible mussel

hître (f)
Ostrea edulis; oyster

couinne (f); vanniaeu (f);
califichieaux (mpl); pitonne (f) [small]
Aequipecten opercularis; queen scallop

pétot (m);
Chlamys distorta; hunchback scallop

vanné (m); scallope (f)
Pecten maximus; scallop

p'tit vanné (m); califichieau (m)
Chlamys varia; variegated scallop

hître (f) dé rocque
Anomia ephippium; saddle oyster

coque (f)
Cerastoderma edule; common cockle

manchot (m); brioche (f) [Trinity];
chuchot (m); râseux (m)
Ensis spp.
Solen marginatus; razorfish

seiche (f); crépie (f) [less common term]
Sepia officinalis; cuttlebone

cônet (m); crépie (f); encônet (f)
Loligo vulgaris; squid (pen)

peurve (f); pieuvre (f); pièvre (f) [L'Etacq]
Octopus vulgaris; octopus

Starfish & Urchins

êtaile (f); êtaile (f) dé mé
Starfish (general term)

tenvrile (f) à la tchilieuvre;
tenvrile (f) dé mé à la tchilieuvre
Ophiura albida; white brittlestar

hérisson (m) d'mé
Paracentrotus lividus; sea urchin
Psammechinus miliaris; purple-tip urchin

nièr brun j'va (m); oeu (m) d'putain
Spatangus purpureus; heart-shaped urchin

Sea Squirts

**câsatchi (m); êclyich'rêsse (f);
êclyich'rêsse (f) dé mé**
Sea squirt (general term)

êclyich'rêsse (f) en lampion
Clavelina lepadiformis; light bulb sea squirt

êclyich'rêsse (f) en grouaîsile
Dendrodoa grossularia; baked bean ascidian

êclyich'rêsse (f) en êtaile
Botryllus schlosseri; star squirt

rouoge êclyich'rêsse (f)
Botrylloides leachi; chain ascidian

Fish

rétchîn (m); cheurque (m)
Lamna nasus; porbeagle shark

**èrnard (m); èrnaûd (m);
r'nard (m); r'naûd (m)**
Alopias vulpinus; thresher shark

**tchian (m); mataud (m) [east];
p'tite rousse (f); rousset (m) [St Ouen];
p'tit rousset (m) [east]**
Scyliorhinus canicula; dogfish

tchian (m) d'mé; rousse (f); roussé (m)
Scyliorhinus stellaris; greater spotted dogfish

haû (m); tchian (m) d'mé
Galeorhinus galeus; tope

démouaîselle (f)
Mustelus asterias; smooth hound

brotchet (m)
Squalus acanthias; spur dog

**ange (f); ange (m) dé mé;
mouaingne (m);
mouaine (m); violon (m)**
Squatina squatina; angel shark

**bourse (f); bourse (f) au dgiâbl'ye;
chiviéthe (f) à bras; crapaud (m) d'mé**
Raja spp.; mermaid's purse

cârrée (f); raie (f)
Raja spp.; ray

**dravan (m) [large]; fliaue (f) [east];
fliée (f) [west]**
Dipturus batis; skate

raietelle (f); rêtelle (f)
Raja micoocellata; small-eye ray

**raie (f) à didget; têgrêsse (f);
tigre-raie (f); tîngrelle (f); tîngue-raie (f)**
Dasyatis pastinaca; stingray

esturgeon (m)
Acipenser sturio; sturgeon

andgulle (f) d'ieau douoche
Anguilla anguilla; eel

**congre (f); andgulle (f); filot (m) [small];
fîlerêsse (f) [small]; fouet (m) [small]**
Conger conger; conger eel

flyie (f)
Alosa alosa; allis shad

flyînte (f); minister (m)
Alosa fallax; jack herring

héthan (m); v'nîse (f) [juvenile]
Clupea harengus; herring

sardinne (f)
Sardina pilchardus; sardine

**êprot (m); esprots (mpl); p'tit hethan (m);
v'nîse (f) [juvenile]**
Sprattus sprattus; sprat

anchouais (m)
Engraulis encrasicolus; anchovy

Jèrriais Species Names

saumon (m)
Salmo salar; salmon

èrnard (m); r'nard (m)
Salmo trutta; sea trout

**paîsson (m) à pouchettes;
raînotte (f) dé mé; scolpîn (m)**
Lophius piscatorius; angler-fish

p'tite louoche (f)
Ciliata mustela; 5-bearded rockling

**mouothue (f); laûdgi (m) [male];
mouorue (f); mouosue (f)**
Gadus morhua; cod

grande louoche (f); r'naud (f)
Gaidropsarus vulgaris; 3-bearded rockling

héthique (f)
Melanogrammus aeglefinus; haddock

**lieu (m) [commonest term]; bideau (m);
gris lieu (m); lieutîn (m) [small];
liotîn (m) [small]**
Merlangius merlangus; whiting

lîn (m)
Molva molva; ling

lieu (m); vidan (m); jaune lieu (m)
Pollachius pollachius; pollack whiting

colîn (m); nièr lieu (m)
Pollachius virens; coalfish

bouothé (m); fliabeu (m)
Trisopterus luscus; bib

**bouosé (m); tabûle (f) [St Ouen];
fliabeu (m) [St Jean]**
Trisopterus minutus; poor cod

mèrluche (f)
Merluccius merluccius; hake

orfi (m); horfi (m)
Belone belone; garfish

grasdos (m)
Atherina presbyter; sand smelt

dorée (f); douothée (m); Jean-Doré (m)
Zeus faber; John Dory

êpinnoche (f)
Spinachia spinachia; sea stickleback

j'va (m) d'mé
Hippocampus hippocampus; seahorse

longnez (m)
Syngnathus spp.; pipefish
Nerophis lumbriciformis; worm pipefish

gronnard (m)
Triglidae; gurnard species

rouoget (m)
Aspitrigla cuculus; red gurnard

grondîn (m)
Eutrigla gurnardus; gurnet

cabot (m) du dgiâbl'ye; grondîn (m)
Myoxocephalus scorpius; short-spined sea
scorpion

crapaud (m) d'mé
Taurulus bubalis; long-spined sea scorpion

**poule (f) dg'ieau; paffot (m);
tambour (m) [St Martin, Trinity]**
Cyclopterus lumpus; lumpsucker

pèrche (f)
Epinephelus marginatus; dusky grouper

bar (m); bâsse (f)
Dicentrarchus labrax; bass

cârré (m)
Trachurus trachurus; horse-mackerel

rouoge brême (f); sarde (f)
Abramis brama; bream

brême sarde (f); sarde (f) [small]
Pagellus bogaraveo; blackspot seabream

néthe brême (f); bliue brême (f)
Spondyliosoma cantharus; black seabream

rouoge mulet (m)
Mullus surmuletus; red mullet

mulet (m); cornelle (f); mulet lippu (m)
Chelon labrosus; thick-lipped mullet

**ouothillard (m); gris mulet (m);
mulet porc (m)**
Liza ramada; grey mullet

co (m) d'la Rocque; co (m) d'rotchi
Centrolabrus exoletus; rock cock

chânaise (f); trie (f); pêtot (m)
Symphodus melops; corkwing wrasse

**cornelle (f); pèrlé (m); pielé (m); vra (m);
êpîle (m) [St Ouen]; couotheux (m) [large]**
Labrus bergylta; ballan wrasse

**coucou (m); râbi (m); démouaîselle (f);
chânaise (f) [female]**
Labrus mixtus; cuckoo wrasse

zèbre (m)
Trachinus draco; greater weever

**sarde (f); vithelîn (m) [St Ouen];
viselun [Grouville]**
Echiichthys vipera; lesser weever

cabot (m)
Lipophrys pholis; shanny

gris cabot (m)
Gobius paganellus; rock goby

nièr cabot (m)
Gobius niger; black goby
70

**rouoge cabot (m); co (m) journieaux;
co (m)**
Parablennius gattorugine; tompot blenny

douoche (f); tchilieuvre (f) dé mé
Pholis gunnellus; butterfish

**lanchon (m); louoche (f);
touoche (f) [female]**
Ammodytes spp.; sand eel

rouoge lanchon (m); p'tit lanchon (m)
Ammodytes tobianus; lesser sandeel

vèrt lanchon (m)
Hyperoplus lanceolatus; greater sandeel

cabot (m) volant
Callionymus lyra; dragonet

célérîn (m); manchouette (f) [east]
Gobius niger; black goby

têtu (m)
Pomatoschistus spp.; sand gobies

maqu'sé (m); maqu'thé (m)
Scomber scombrus; mackerel

thon (m)
Thunnus thynnus; tunny-fish

cat (m); catte (f) [east]
A general term for all species of flatfish.

fiandre (f); fliandre (f)
Phrynorhombus norvegicus; Norwegian
topknot

**turbot (m) [west];
teurbot (m) [east]**
Psetta maxima; turbot

brille (f)
Scophthalmus rhombus; brill

sole (f) dé rotchi
Zeugopterus punctatus; topknot

fliêtan (m); halibot (m) [east]
Hippoglossus hippoglossus; halibut

pliaie (f)
Pleuronectes platessa; plaice

sole (f)
Solea solea; sole

lînmon sole (m)
Microstomus kitt; lemon sole

solé (m)
Mola mola; sunfish

Seals & Dolphins
**sirène (f); syraine (f);
loup-mathîn (m);
vielle fil'ye (f)**
Halichoerus grypus; grey seal

dauphîn (m)
Tursiops truncatus; bottle-nosed dolphin

pourpais (m); pourpé (m) [east]
Phocoena phocoena; porpoise

Seaweeds
collet (m)
Laminaria digitata; oarweed

**collet (m);
tangon (m) [stalk]**
Saccorhiza polyschides; sugar kelp

ivraie (f)
Halidrys siliquosa; sea oak

crochet (m); crachet (m); vrégîn (m)
Pelvetia canaliculata; channelled wrack
**vrégîn (m); pliat vrai (m); cracot (m);
crochet (m)**
Ascophyllum nodosum; knotted wrack

pliat vrai (m); vrégîn (m); vrégeais (m)
Fucus serratus; serrated wrack

**bédanne (f); pliat vrai (m); vrégîn (m);
vrégeais (m); bédaine (f);
vrai (m) à cliouques**
Fucus vesiculosus; bladder wrack

lachon (m); lachet (m)
Himanthalia elongata; thongweed

mousse (f)
Corallina officinalis; sea moss

**vèrdidget (m); litchet (m);
moussinn'nie (f)**
Chondrus crispus; carrageen

cliouque (f); cliaque (f)
Ulva intestinalis; gutweed
Ulva lactuca; sea lettuce
Porphyra purpurea; laver

Seagrasses
hèrbi (m); plîse (f)
Zostera marina; eelgrass; seagrass

Les Faucheurs

Further Reading

Channel Islands

Chambers, P., 2008. *Channel Island Marine Molluscs*. Charonia Media.

Daly, S., 1998. *Marine Life of the Channel Islands*. Kingdom.

Le Sueur, R., 1967. *The Marine Fishes of Jersey*. La Société Jersiaise.

McIlwee, K., 2015. *Jersey Scuba Explorers' Guide*. Barnes Publishing.

Sinel, J., 1906. *An Outline of the Natural History of Our Shores*. Swan Sonnenschein.

General Identification Guides

Audibert, C. & Delemarre, J.-L., 2009. *Guides des Coquillages de France: Atlantique et Manche*. Belin.

Bunker, F., *et al*, 2012. *Seaweeds of Britain and Ireland*. Wild Nature Press.

Hayward, P., *et al*, 1996. *Seashore Life of Britain and Europe*. Collins Pocket Guide.

Hayward, P. & Ryland, J., 2005. *Handbook of the Marine Fauna of North-West Europe*. Oxford University Press.

Kay, P. & Dipper, F., 2009. *Marine Fishes of Wales*. Marine Wildlife.

Martin, J., 2011. *Les Invertébrés Marins du Golfe du Gascogne à la Manche Orientale*. Editions Quae.

Picton, B., *et al*, 2007. *Sponges of the British Isles*. Marine Conservation Society. [Available free online]

Porter, J., 2012. *Guide to Bryozoans and Hydroids of Great Britain and Ireland*. Marine Conservation Society.

Sterry, P. & Cleave, A., 2012. *Complete Guide to British Coastal Wildlife*. Collins.

Wood, C., 2005. *Sea Anemones and Corals of Britain and Ireland*. Marine Conservation Society.

Wood, C., 2007. *Observer's Guide to Marine Life of Britain and Ireland*. Marine Conservation Society.

Seymour Tower

Further Reading, Credits

Acknowledgements

We thank the following organisations for their assistance with the production of this guide: Nedbank Private Wealth, Société Jersiaise; Jersey Seasearch; L'Office du Jèrriais; Environment Department (States of Jersey); Jersey Aquatic Discovery; Little Feet Environment; Sea Shepherd; Jersey Seafaris; Bouley Bay Dive Centre.

We also thank the following individuals for their help with this and other marine biology projects: Samantha Andrews; Louise Bennett-Jones; Francis Binney; Samantha Blampied; Simon Bossy; Bertram Brée; Paul Chambers; John Clarke; Judy Collins; Fiona Crouch; Sabina Danzer; Charles David; Bob de la Haye; Andy Farmer; Judith Freeman; Anne Haden; Derek Hairon; Ed and Annie Hibbs; Courtney Huisman; Gareth Jeffreys; Geraint Jennings; Ashley Johnson; Nicholas Jouault; Phillip Langlois; Tony Legg; Loftur Loftsson; Roger Long; Richard Lord; Caroline Leach; Anya Martins; John McLaughlin; Kevin and Beverly McIlwee; Neil Molyneux; Greg Morel; Kirsten Morel; John Noel; Sadie Norman; Jon Rault; Aimee Reading; Tony Scott-Warren; Jon Shrives; Jillian Smith; Mike Smith; Frankie Stammers; Andrew Syvret; Pauline Syvret; Deryk Tolman; Bob and Jill Tompkins; Trudie Trox-Hairon; Kirk Truscott; Geoff Walker; Marion Walton; Chris Wood; Tim Wright

Picture Credits

Below is a list of the photographers who kindly agreed to donate their images to this book and to whom we are extremely grateful. Of the 554 photographs reproduced, all but eight were taken in Jersey.

TL = top left; TR = top right; ML = middle left; MR = middle right; BL = bottom left;

BR = bottom right; (i) = inset image; SoJ = States of Jersey

8-TL(i):	Gareth Jeffreys	23-TL:	Gareth Jeffreys	44-ML:	Kirk Truscott
8-BL:	Kevin McIlwee	23-MR:	Kate Binney	44-BR:	Kirk Truscott
9-ML:	Gareth Jeffreys	23-ML:	Simon Butler	45-ML:	Gareth Jeffreys
9-BL;	Chris Wood	25-MR:	Gareth Jeffreys	45-BR:	Bob de la Haye
10-ML:	Gareth Jeffreys	25-BR:	Kate Binney	45-TL:	Kevin McIlwee
10-BL:	Gareth Jeffreys	28-TR(i):	Gareth Jeffreys	45-BR:	Hans Hillewaert
10-ML(i):	Sabina Danzer	28-ML(i):	Gareth Jeffreys	45-TR:	Kevin McIlwee
11-TL:	Gareth Jeffreys	31-TR(i):	Bob Tompkins	45-MR:	Kirk Truscott
11-ML(i):	Sabina Danzer	31-ML:	Gareth Jeffreys	45-MR(i):	Marine Resources (SoJ)
15-TR:	Auguste le Roux	33-TR:	Gareth Jeffreys	46-TR:	Gareth Jeffreys
15-BL:	Auguste le Roux	38-TR:	Kevin McIlwee	46-ML:	Gareth Jeffreys
15-BR(i):	Marine Resources (SoJ)	39-TR:	Kevin McIlwee	46-BL:	Gareth Jeffreys
15-BR:	Marine Resources (SoJ)	40-TR:	Gareth Jeffreys	46-MR:	Kevin McIlwee
17-BR:	Gareth Jeffreys	40-ML(i):	Gareth Jeffreys	47-TR:	Citron
18-TL:	Gareth Jeffreys	41-TL:	Gareth Jeffreys	47-BR:	Gareth Jeffreys
18-BR:	Gareth Jeffreys	41-BL(i):	Gareth Jeffreys	47-BR(i):	Jersey Seafaris
18-TL(i):	Marine Resources (SoJ)	42-MR:	Chris Wood	47-BL:	Jersey Seafaris
19-MR:	Gareth Jeffreys	42-BL:	Bob de la Haye	57-MR(i):	Gareth Jeffreys
20-ML(i):	Gareth Jeffreys	42-BR:	Chris Wood	57-BR:	Gareth Jeffreys
21-TR:	Gareth Jeffreys	43-BR:	David Tipping	61-BL(i):	Gareth Jeffreys
21-TL:	Gareth Jeffreys	43-TL(i):	Gareth Jeffreys	63-MR(i):	Marine Resources (SoJ)
21-ML:	Kate Binney	43-MR(i):	Gronco	63-MR:	Marine Resources (SoJ)
22-ML(i):	Kevin McIlwee	43-BL:	Kirk Truscott	All other photographs	
23-MR(i):	Francis Binney	43-TR(i):	Matthieu Sontag	by Paul Chambers	

We have made every attempt to identify correctly the species portrayed but errors are always possible. We would be grateful to be informed of any misidentifications.

Index of Scientific Names

Index of Scientific Names

Index of Scientific Names

Les Ecréhous

Index of Common Names

Index of Common Names

Index of Common Names

There have been 3,216 marine species recorded by the Société Jersiaise from the Channel Islands which is probably about half the actual total. A tally of these (by their phylum) is provided below.

Phylum	Common Name	No. of Species
PROTOCTISTA	Protists (general)	4 species
SARCOMASTIGOPHORA	Sarcomastigophorids	5 species
DINOMASTIGOTA	Dinoflagellates	6 species
CILIOPHORA	Ciliates	18 species
FORAMINIFERA	Foraminiferids	140 species
PORIFERA	Sponges	115 species
CNIDARIA	Jellyfish, Corals, Anemones	157 species
CTENOPHORA	Comb Jellies	2 species
PLATYHELMINTHES	Flat Worms	9 species
ACOELOMORPHA	Marine Worms	2 species
NEMERTEA	Ribbon Worms	30 species
ROTIFERA	Rotiferids	2 species
PRIAPULIDA	Priapulid Worms	2 species
ENTOPROCTA	Entorpocts	5 species
CHAETOGNATHA	Arrow Worms	2 species
SIPUNCULA	Spoon Worms	6 species
ANNELIDA	Bristle Worms	219 species
CHELICERATA	Sea Spiders, Mites	38 species
MYRIAPODA	Centipedes	3 species
LOWER CRUSTACEA	Barnacles, Ostracods	127 species
HIGHER CRUSTACEA	Crabs, Prawns, Lobsters	409 species
HEXAPODA	Insects	11 species
TARDIGRADA	Tardigrades	8 species
MOLLUSCA	Snails, Clams, Sea Slugs	515 species
BRACHIOPODA	Brachiopods	3 species
PHORONIDA	Phoronids	1 species
BRYOZOA	Sea Mats	160 species
ECHINODERMATA	Starfish, Sea Urchins	56 species
HEMICHORDATA	Hemichordates	2 species
CHORDATA	Sea Squirts	84 species
CHORDATA	Lancelet	1 species
CHORDATA	Fishes	191 species
CHORDATA	Turtles	4 species
CHORDATA	Dolphins, Whales, Seals	15 species
BACILLARIOPHYTA	Diatoms	251 species
OCHROPHYTA	Orchrophytes	146 species
CRYPTOPHYTA	Cryptopytes	1 species
GRACILICUTES	Cyanobacteria	45 species
RHODOPHYTA	Seaweeds	286 species
CHLOROPHYCOTA	Chlorophytes	82 species
ANGIOSPERMOPHYTA	Flowering Plants	3 species
ASCOMYCOTA	Lichens	50 species

www.ingramcontent.com/pod-product-compliance
Lightning Source LLC
Chambersburg PA
CBHW040130270326
41928CB00001B/17